Cell Science

Sally Cowan

Australia • Brazil • Japan • Korea • Mexico • Singapore • Spain • United Kingdom • United States

Cell Science

Text: Sally Cowan
Editor: Ben Haskin
Design: Karen Mayo
Series design: James Lowe
Photo researcher: Lisa Piemonte
Production controller: Adam Bextream
Reprint: Jennifer Foo

Acknowledgements
Thanks to Professor Peter Cowan of the Immunology Research Centre at St Vincent's Hospital, Melbourne, for technical consultation and assistance in writing this book.

The author and publisher would like to acknowledge permission to reproduce material from the following sources:
Auscape/BSIP/Jacopin: p. 11 (bottom); Corbis Australia: p. 18; Getty Images: pp. 1, 16, 23, cover; iStockphoto: p. 10 (top centre); iStockphoto/Jacob Moisin: p. 9 (main); iStockphoto/Wouter van Caspel: p. 13; J.C. Revy/Science Photo Library: p. 21 (top); Jupiterimages Corporation © 2009: p. 12; Newspix/AFP Photo/Treasury Department: p. 17; Photolibrary: pp. 3, 4, 6, 7 (both), 9 (inset), 10 (top left), 10 (top right), 10 (bottom left), 10 (bottom centre), 10 (bottom right), 11 (top), 14, 15 (both), 19, 20, 21 (bottom), 22; Shutterstock/Ivan Cholakov: back cover; Shutterstock/Jan Kaliciak: p. 5; Shutterstock/Katrina Leigh: p. 8.

Fast Forward Independent Texts
Level 23

For product information and technology assistance,
in Australia call 1300 790 853;
in New Zealand call 0508 635 766

For permission to use material from this text or product,
please email **aust.permissions@cengage.com**

ISBN 978 0 17 018102 0
ISBN 978 0 17 017899 0 (set)

Cengage Learning Australia
Level 7, 80 Dorcas Street
South Melbourne, Victoria Australia 3205

Cengage Learning New Zealand
Unit 4B Rosedale Office Park
331 Rosedale Road, Albany, North Shore NZ 0632

For learning solutions, visit **cengage.com.au**

Printed in Australia by Ligare Pty Ltd
2 3 4 5 6 7 8 23 22 21 20 19

Cell Science

Sally Cowan

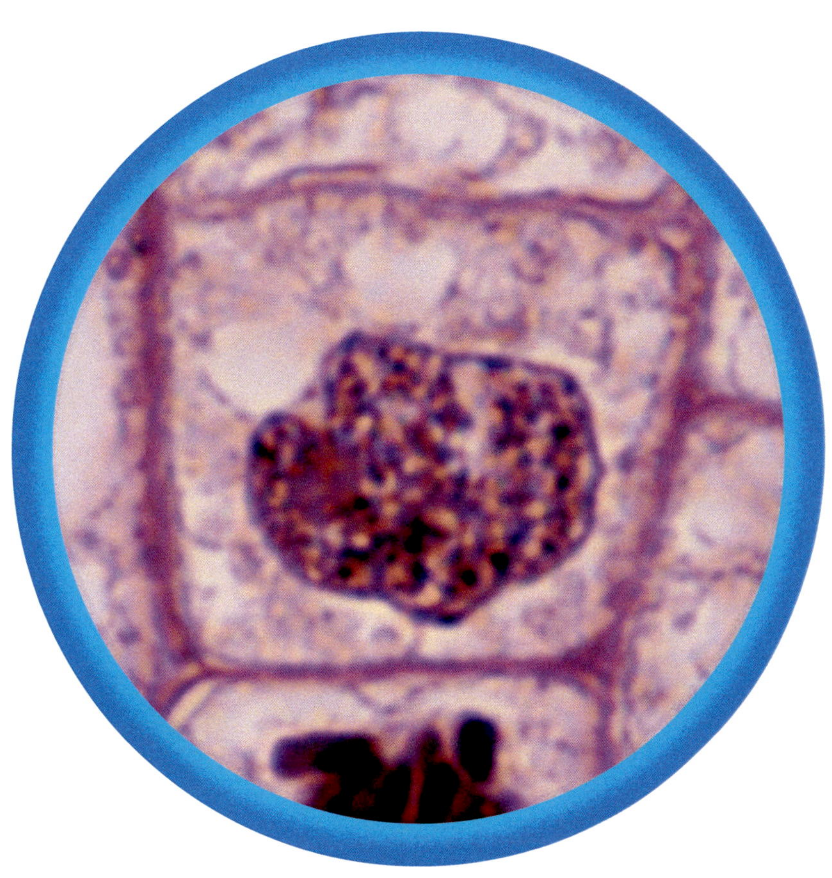

Contents

Cells and Living Things

Cells are the most basic form of life.
They are so tiny that they can only be seen with a microscope.

All living things are made up of either one cell or many cells.
A cell **divides** to make two identical "daughter" cells.
This is how living things grow.

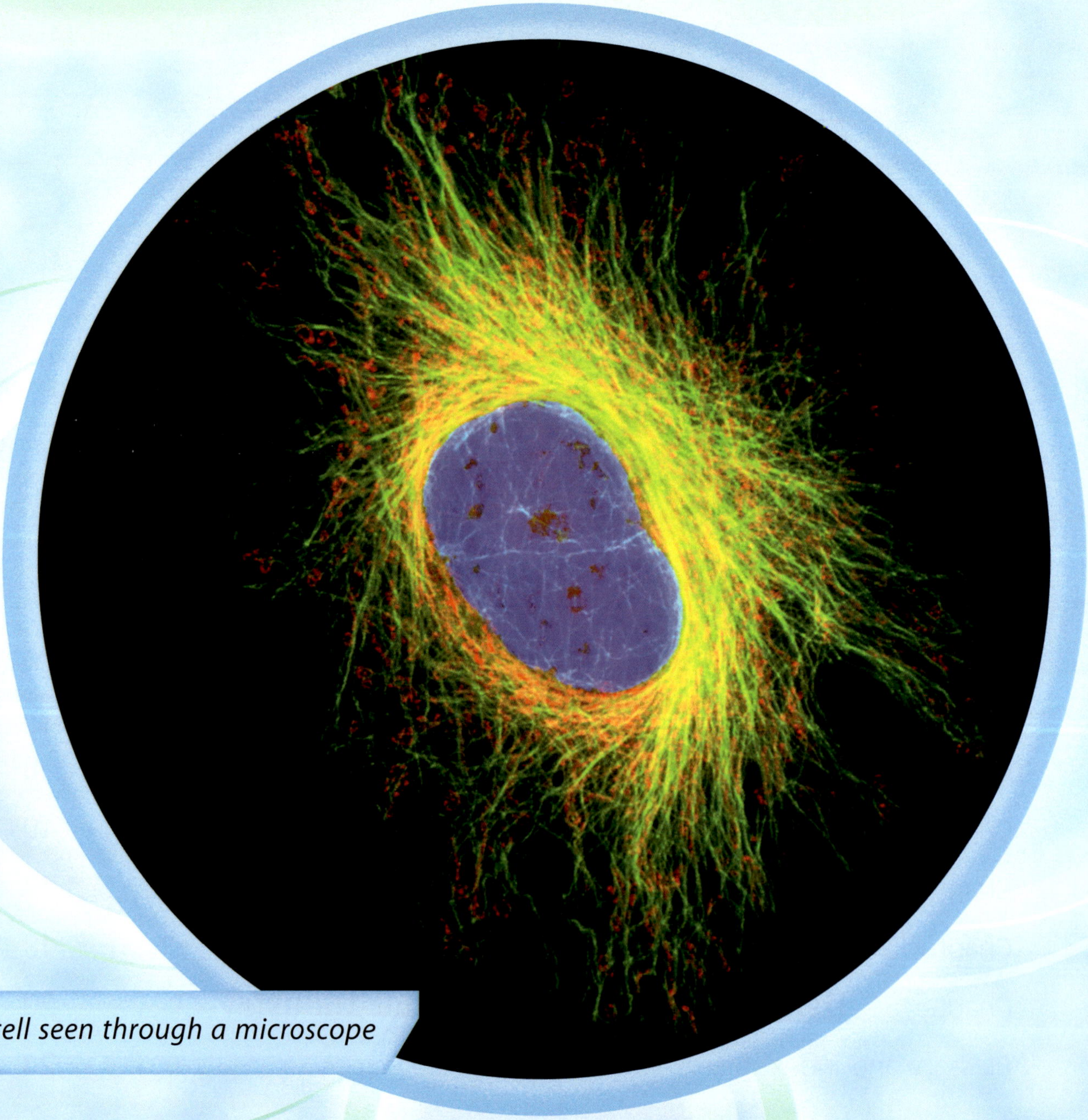

a cell seen through a microscope

Although cells are tiny, a lot goes on inside them. Chemical processes take place that allow cells to live and do their job.

A substance called **DNA** is inside each cell. The DNA contains a set of instructions called genes. Genes determine what an **organism** is and how it works.

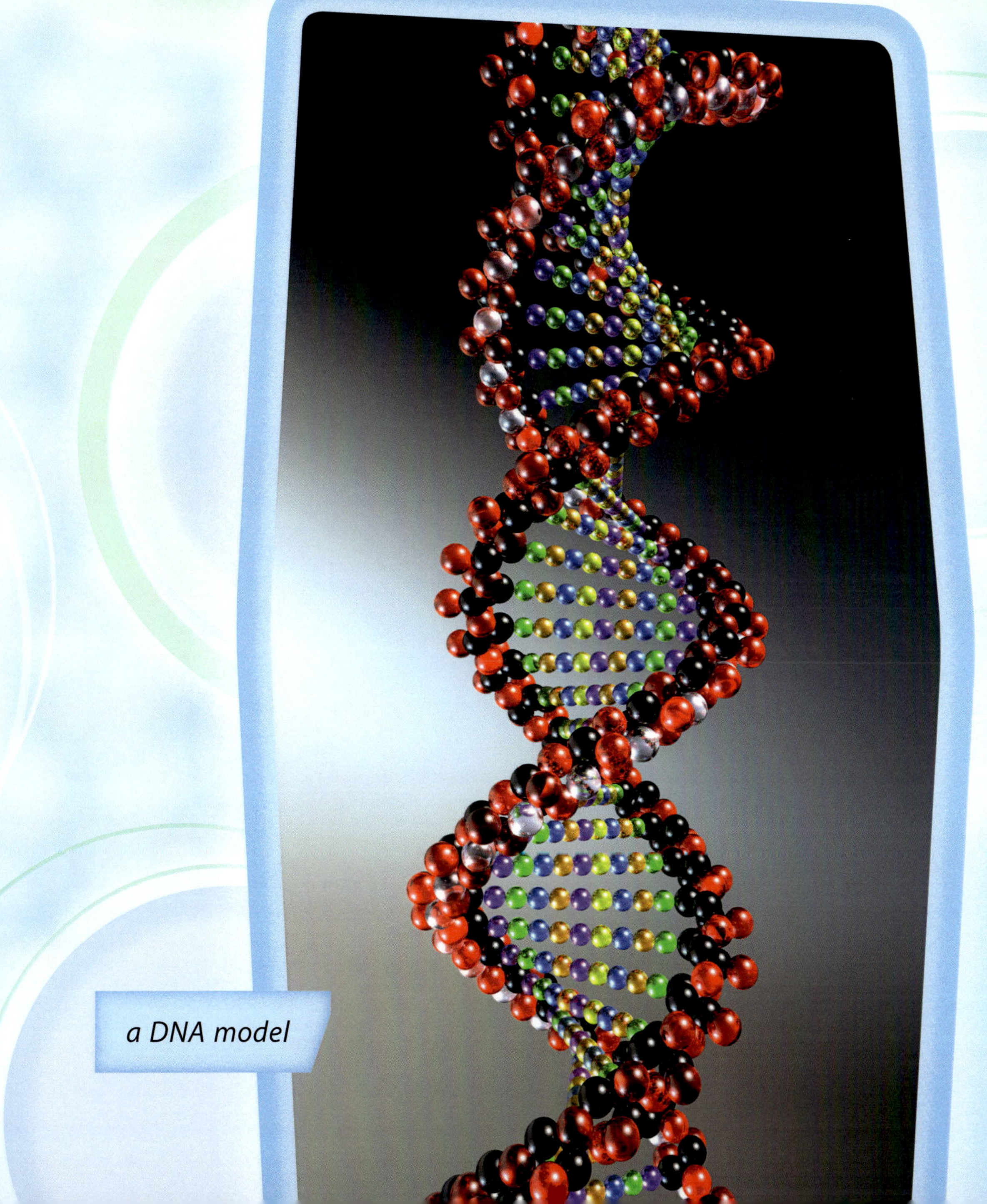

a DNA model

Single-Cell Organisms

Simple living things, like bacteria, are made of a single cell.

When a **bacterium reproduces**, it divides to form two daughter cells.
Some bacteria can divide in under ten minutes.
This is how bacteria can spread quickly.

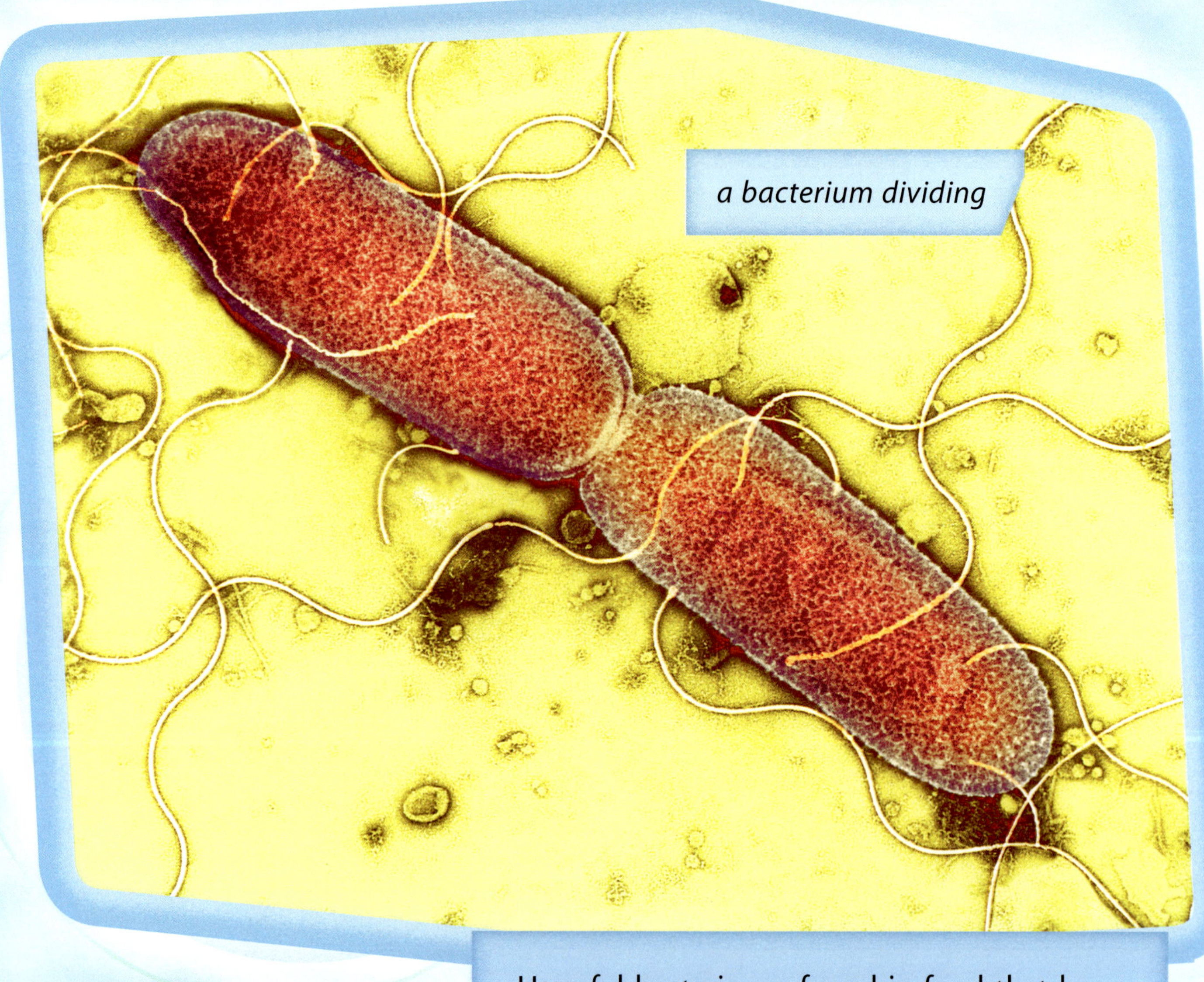

a bacterium dividing

Harmful bacteria are found in food that has not been prepared or stored properly.
If eaten, the bacteria can make people sick.

Multi-Cell Organisms

Animals and plants are made of many types of cells that do different things.
The cells of animals and plants have very different **structures**.

Plant cells have hard walls that stay the same shape.
Animal cells have **flexible** walls that allow them to change shape.
This is one reason why animals and plants are so different.

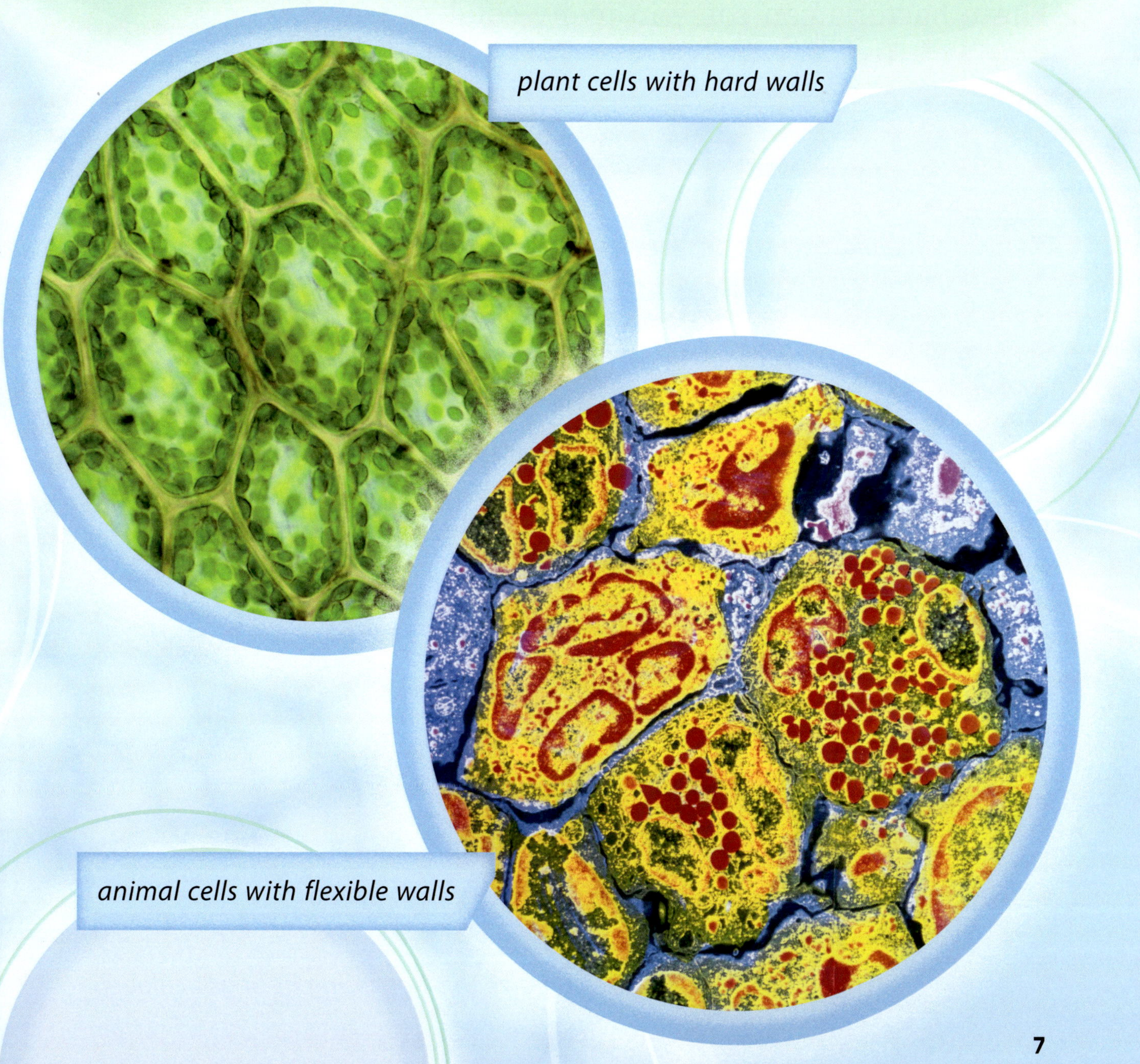

plant cells with hard walls

animal cells with flexible walls

CHAPTER 2

Extreme Single Cells

Scientists believe that single-cell organisms, like bacteria, were the first living things on Earth.
Different types of bacteria have adapted to live in most places on Earth.

Some bacteria live at very high temperatures.
They can even live in hot springs and mud pools.
These bacteria can come in many colours.

Bacteria give this hot spring its bright colours.

Other bacteria live at very low temperatures.
They can be found in icy places such as Antarctica, on mountains and in glaciers.
They contain special chemicals that stop them from freezing.

anti-freeze

Some cold-loving bacteria contain anti-freeze chemicals, a little like those used in cars.

Human Cells

Humans have **trillions** of cells.
The cells work together to keep the body alive and well.
They come in many different shapes and sizes and do different things.

bone cell

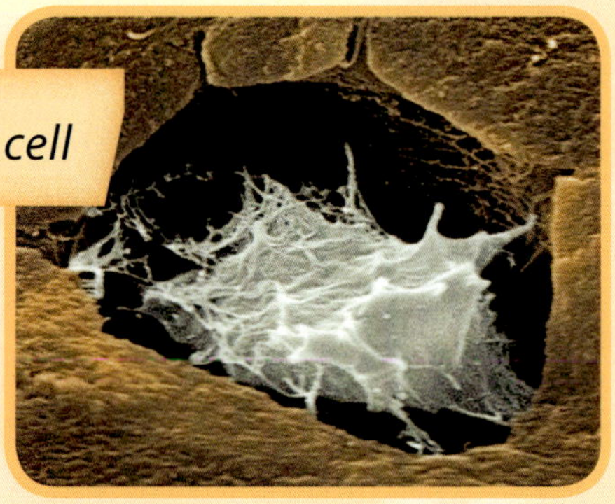

white blood cell

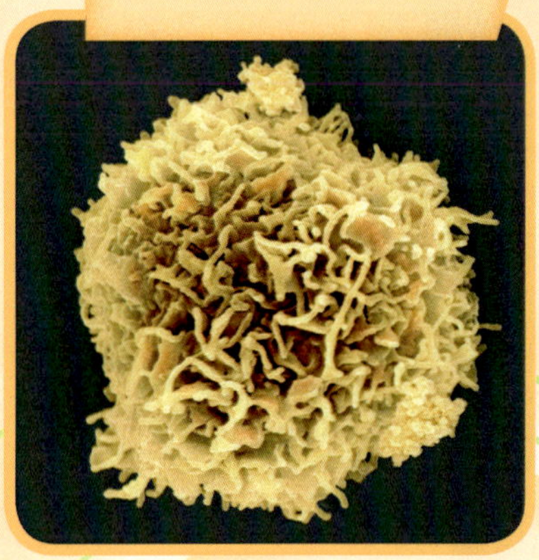

liver cells

sperm cell

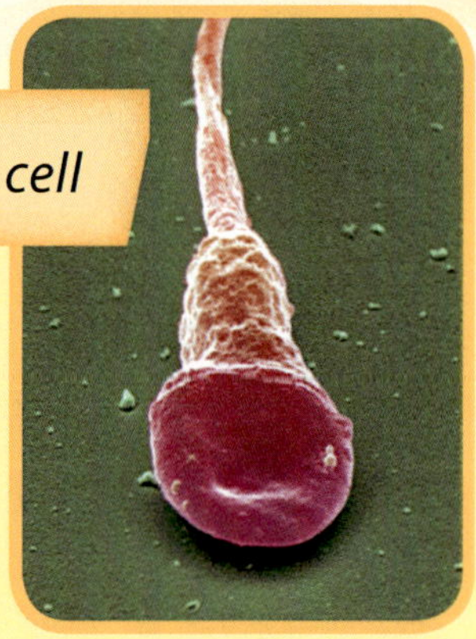

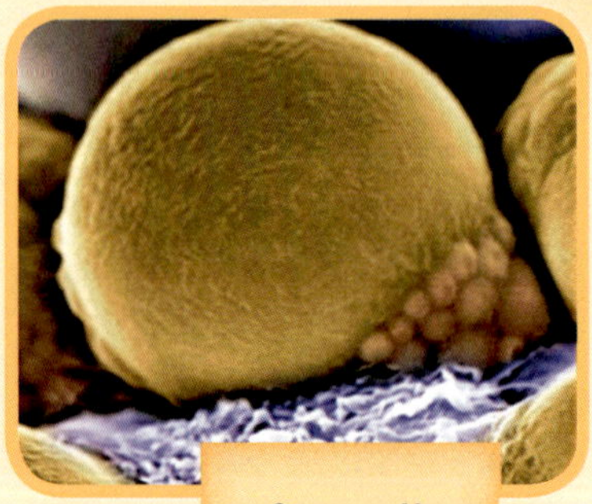

red blood cell

fat cell

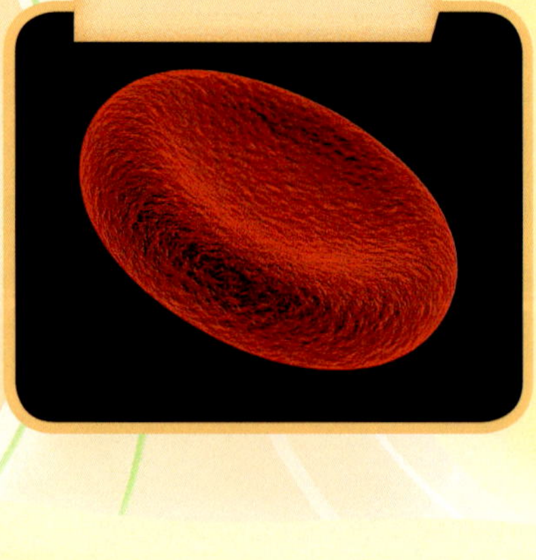

All cells need oxygen to live.
Red blood cells, shaped like discs, float in the blood.
They take oxygen to the other cells in the body.

Muscle cells are long and thin.
They pull together to move parts of the body.

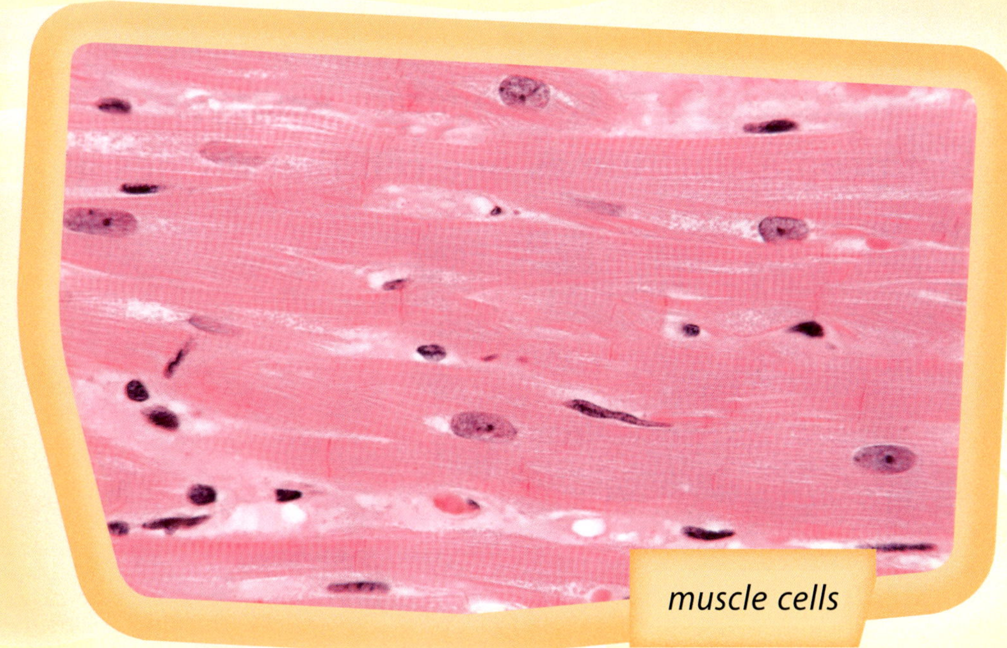

muscle cells

Skin cells are flat and closely packed.
They form a layer to protect the body.

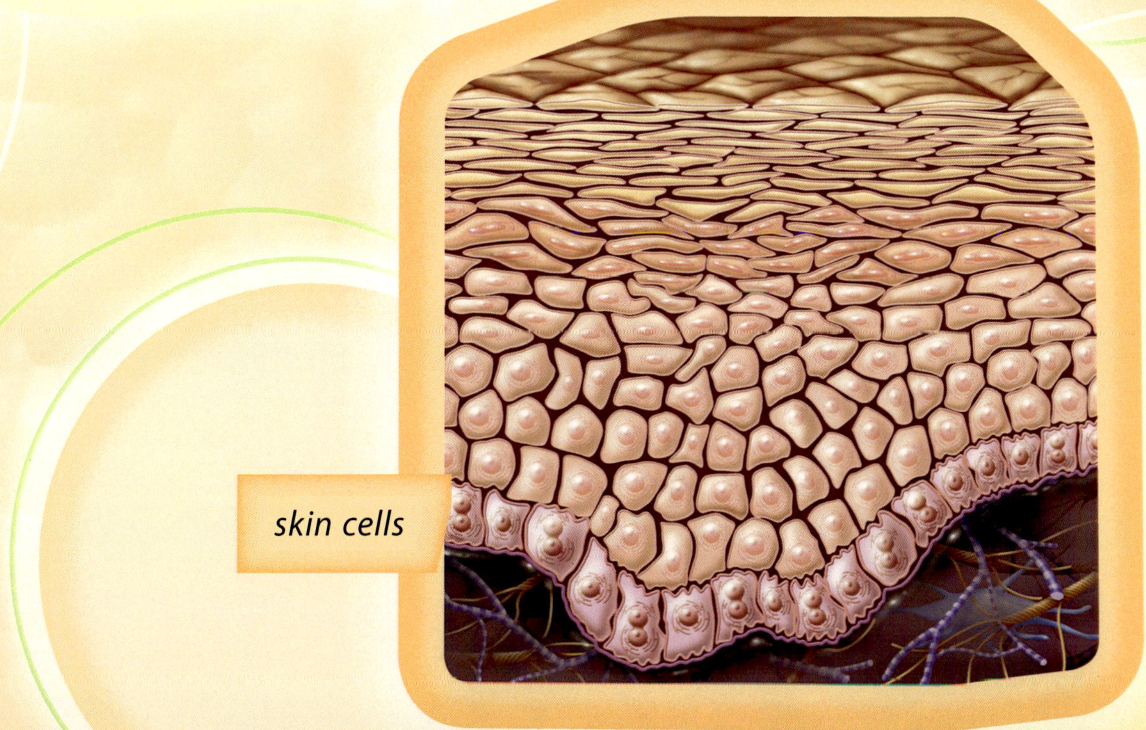

skin cells

DNA

All people have very similar DNA.
Only identical twins have exactly the same DNA.
This is because identical twins start life from one cell that divides to form two separate **embryos**.
Most people start life as one cell that divides and forms one embryo.

identical twins

Genes

Genes are passed on in the cells from parents to children. Children have a mix of genes from their mother and father.

The genes determine what people will look like and how their bodies will work.
But most scientists believe that the environment also influences how people turn out.

Genes are passed down within families. That is why family members often look, sound and even act alike.

Cells Out of Control

In humans and other **multi-cell** organisms, each type of cell lives for a fixed length of time.

Every second, many cells inside the human body die and are replaced with new ones.

Skin cells live for about four weeks, while red blood cells live for four months.

The immune system, which protects the body from disease, has some cells that live for years.

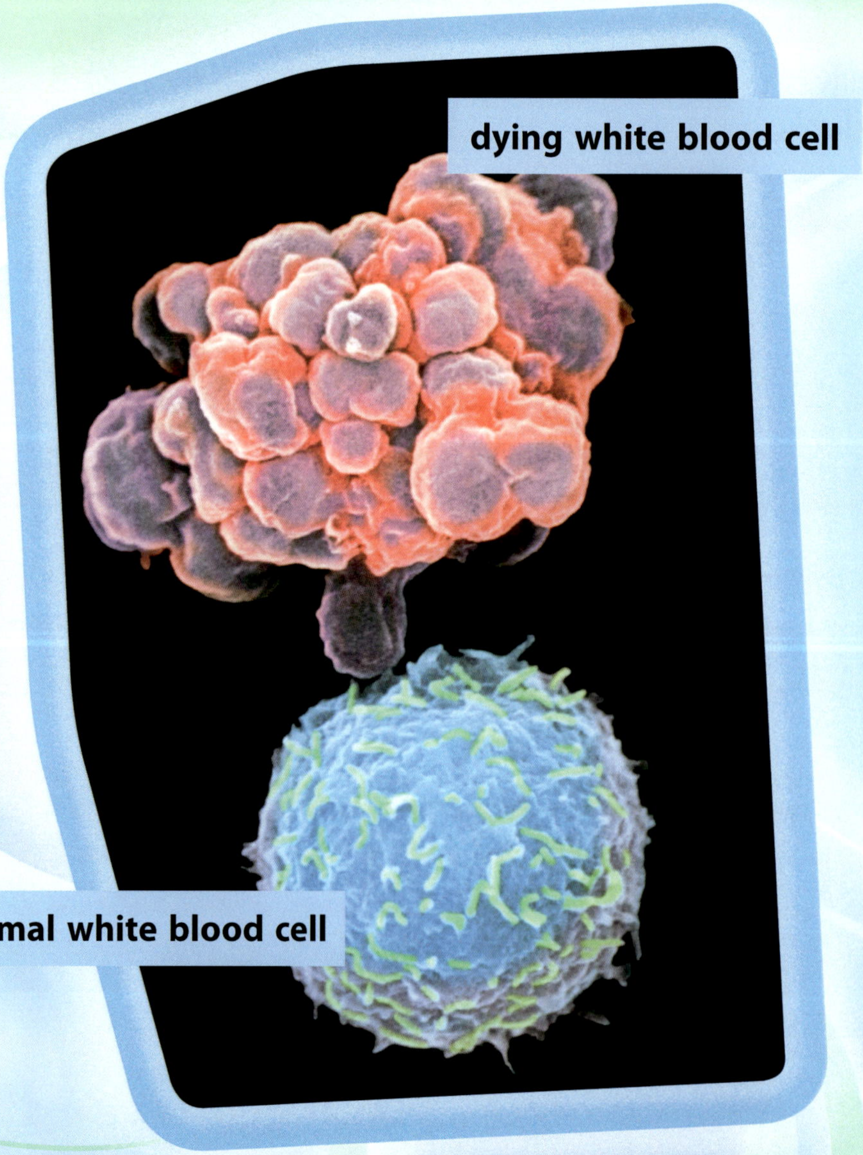

dying white blood cell

normal white blood cell

Sometimes cells do not die when they should, and divide out of control.

Most of the time, the immune system kills off these cells before they cause problems.

Sometimes they are not killed and cause a disease called cancer.

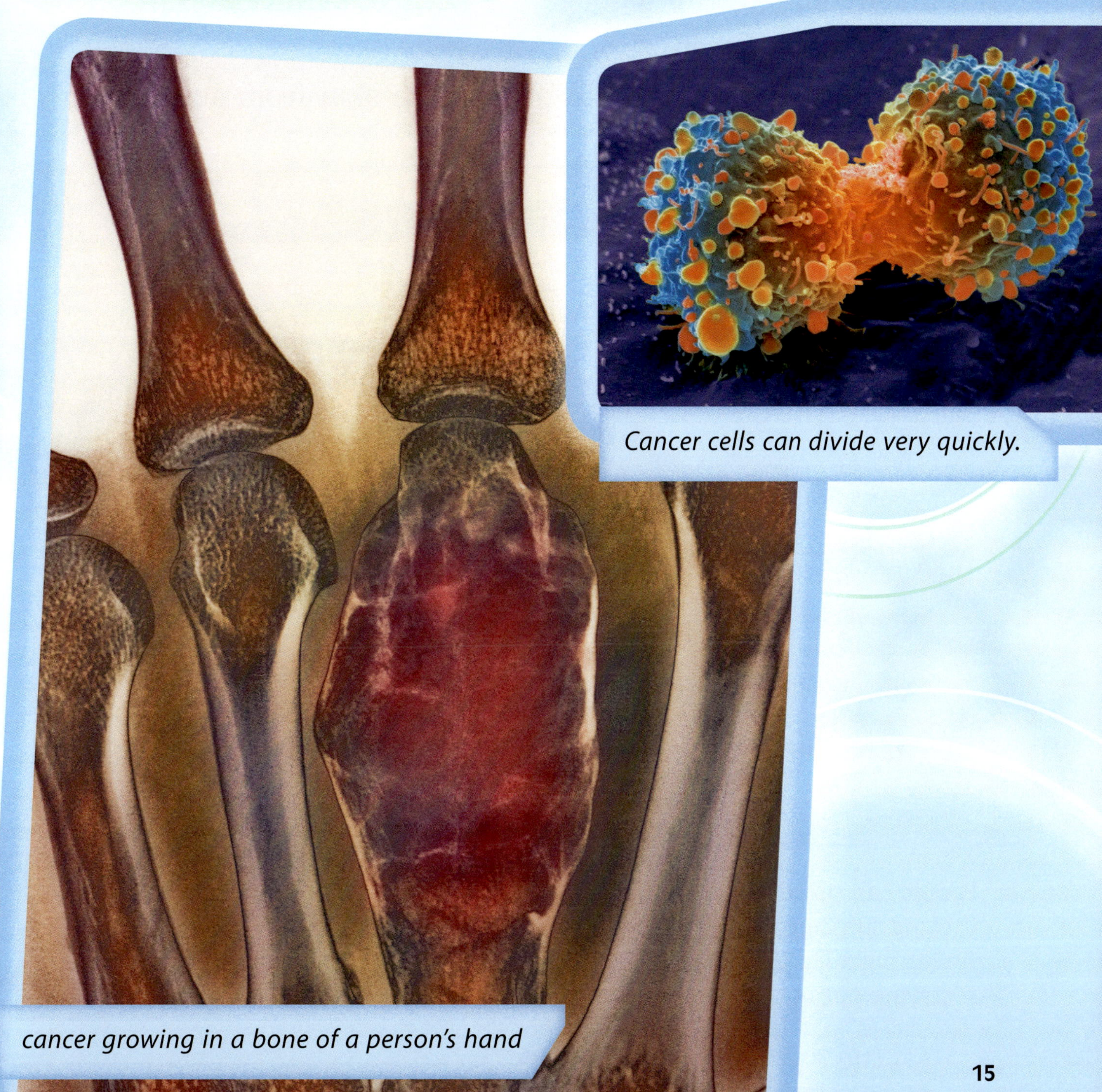

Cancer cells can divide very quickly.

cancer growing in a bone of a person's hand

Most cancers are caused by changes to a cell's DNA. Cells must copy their DNA when they divide. Sometimes the DNA is not copied properly. Other times, a problem in a person's genes may change their DNA and give them a higher chance of getting cancer.

There are also some environmental factors that change DNA. Too much sunlight can damage the DNA of skin cells and cause skin cancer. That is why it is important to protect the skin from sunlight.

People can protect their skin by

- using sunscreen
- wearing a hat and a long-sleeved shirt
- staying out of the sun in the middle of the day.

Cigarette smoke can damage the DNA of cells in the lungs and cause lung cancer.

Many countries have health campaigns to inform people of the dangers of cigarette smoking and to help them stop smoking.

The History of Cell Science

Robert Hooke was the first person to discover cells. In 1665, he examined thin pieces of cork under a microscope and saw lots of little box shapes joined together. He called them "cells".

Better microscopes allowed scientists to examine animal cells. In 1839, two German scientists suggested that living things were made up of one cell or many cells that worked together.

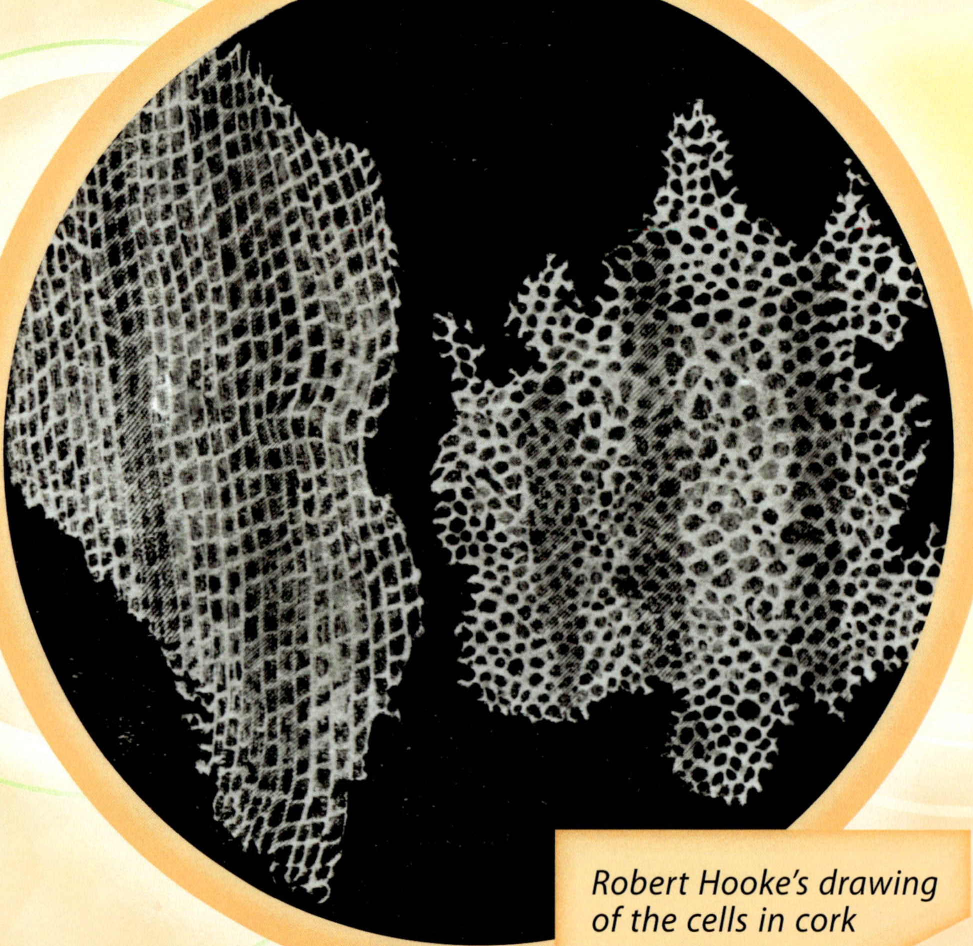

Robert Hooke's drawing of the cells in cork

In 1944, three Canadian scientists worked out that DNA held a cell's instructions.

Then, in 1953, James Watson and Francis Crick discovered the structure of DNA.

Watson and Crick with their model of DNA

Since then, scientists have discovered that human DNA contains 30000 to 40000 genes.

Using Cells in Medicine and Industry

Cell science has developed quickly over the last 50 years, and cells are now used in many different ways.

Cell science is used to make some medicines, like antibiotics, which treat infections and diseases.

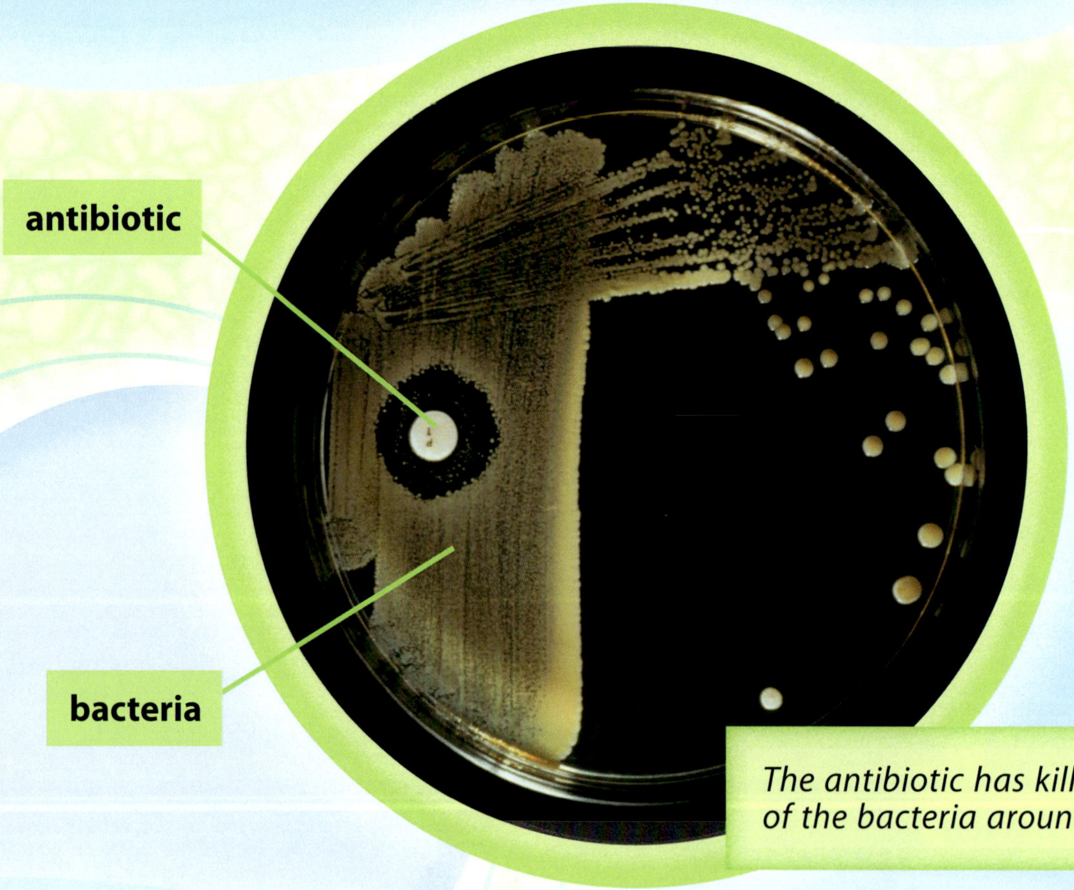

antibiotic

bacteria

The antibiotic has killed some of the bacteria around it.

The DNA from people's cells can be tested to find out if they are likely to develop a disease.
Doctors may be able to stop the disease from developing, or treat it early.

Scientists use skin cells to grow new skin for treating people with burns.

growing skin in a laboratory

Cell science is also used to clean up waste materials. Bacteria break down waste into useful things like gases, which can be kept and used.

This sewage treatment plant uses bacteria to break down sewage into methane, which can be used as a fuel.

The Future of Cell Research

A lot of research is being done into **stem cells**.
Stem cells can develop into any type of cell in the body.

a researcher examining a stem cell through a microscope

At first, scientists thought human embryos were the only source of stem cells.
Many people disagree with using stem cells from embryos.
But researchers can now use stem cells from adults.

At present, stem cells are used to help treat some cancers of the blood.
In the future, they may be used to replace or fix damaged cells that cause other diseases.
Scientists even hope to use stem cells to help people in wheelchairs to walk again.

There is a lot to look forward to in the science of cells.

Glossary

bacterium singular of "bacteria"

divides splits apart

DNA deoxyribonucleic acid; the material inside a cell that determines how the cell will develop

embryos the beginnings of a plant or animal

flexible easily bent or moved

multi-cell made of more than one cell

organism a living thing

reproduces produces offspring

stem cells cells of no set type that can turn into any cells

structures the way things are built, their frameworks

trillions millions of millions; one trillion, or ten to the power of 12, is written 1 000 000 000 000

Index